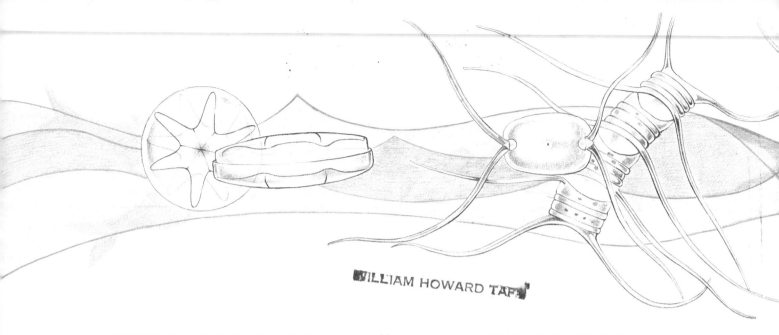

# JULIAN MAY

### illustrations by JEAN ZALLINGER

*Holiday House* • *New York*

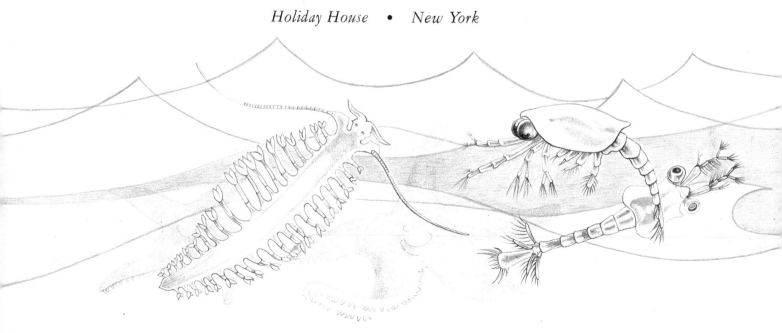

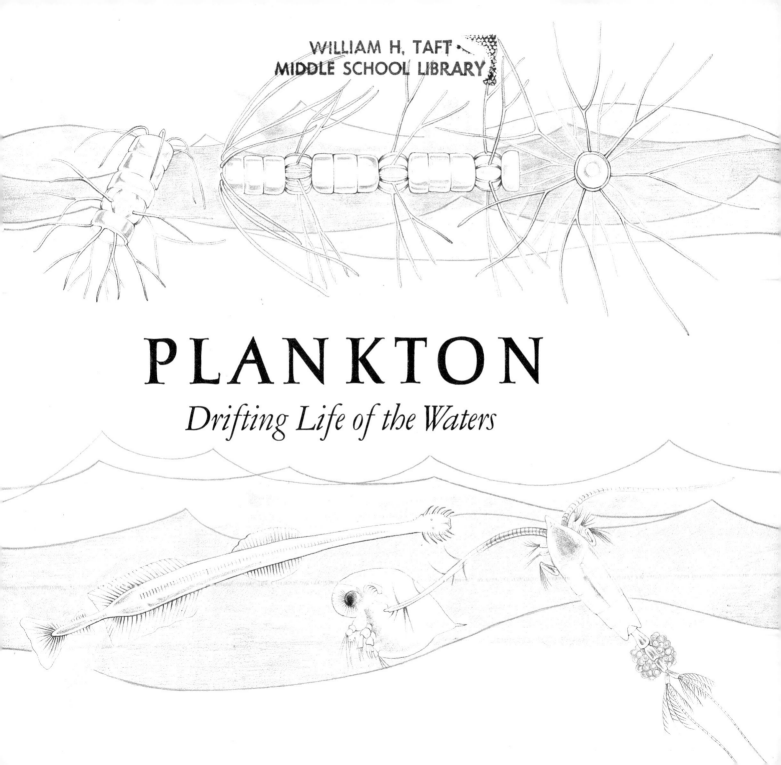

# PLANKTON

*Drifting Life of the Waters*

# PLANKTON

*Drifting Life of the Waters*

Travel in a boat through the open waters, through waters that seem empty — without fish or plants. Drag a cloth bag behind the moving boat. Scrape out some of the brownish stuff you find in the bag and put it into pond water — not tap water. Look at it under a microscope and see . . .

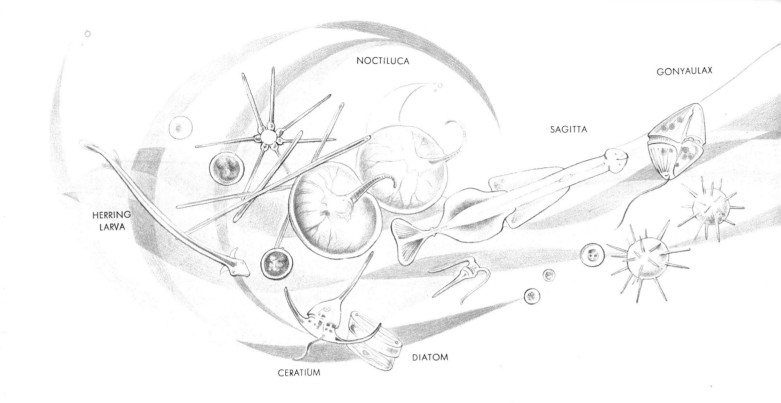

NOCTILUCA

GONYAULAX

SAGITTA

HERRING LARVA

DIATOM

CERATIUM

Tiny animals that look as if they were made of glass,
darting, swimming, drifting, wiggling . . .
Little hunters of rainbow-colored jelly, taking prey
with poison darts . . .
Plants like jewel boxes with a bit of green inside. . . .

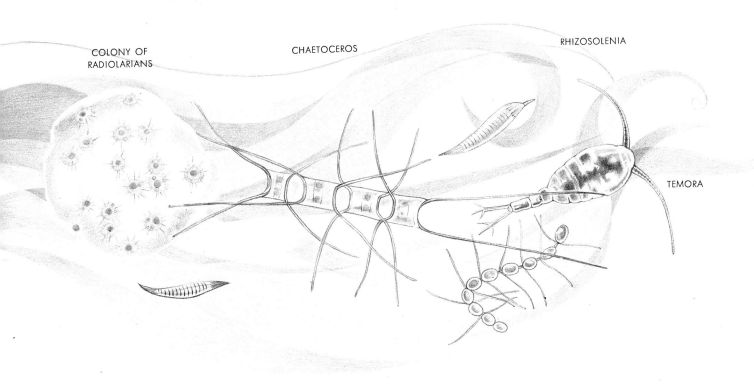

COLONY OF RADIOLARIANS    CHAETOCEROS    RHIZOSOLENIA    TEMORA

They are the *plankton*, plants and animals that drift
through the water, moved by currents and tides.
The word "plankton" means "that which is carried along."
Many of the animals — and even some plants — can swim,
but only weakly. Their own movements do not carry
them far.

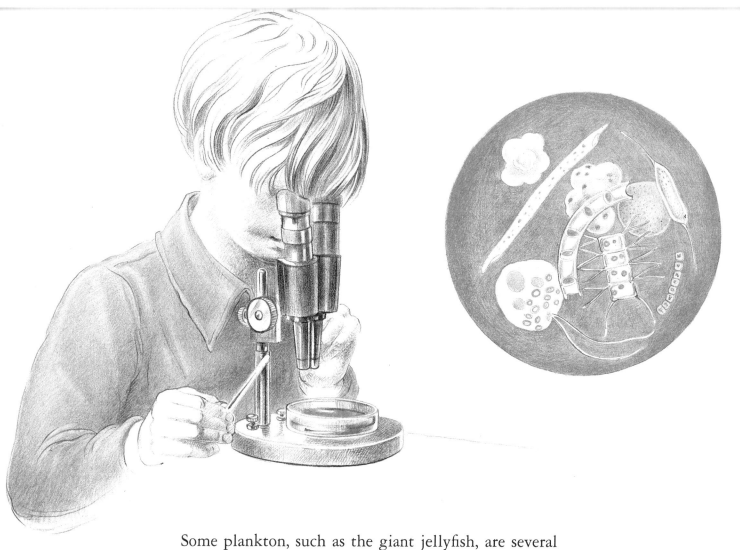

Some plankton, such as the giant jellyfish, are several feet wide. But most of the animals are much, much smaller. They are seen best with a good magnifying glass or a microscope. The plant plankton are invisible, and there are countless numbers of them living in the sea, in lakes, ponds, and other waters.

Even though they are small, the plant plankton are probably the most important living things on earth. They have a green chemical called chlorophyll. With the help of this chemical they can make their own food. They take in water and a gas called carbon dioxide. Using sunlight for energy, they make food and give off oxygen and water as "leftovers."

CARBON DIOXIDE
FROM AIR

OXYGEN
GIVEN
OFF

WATER

MORE WATER
GIVEN OFF

Green plants, when there is light, take in carbon dioxide and breathe out oxygen

Many scientists believe that most of the earth's oxygen was made by plant plankton in the oceans. These bits of plant life have given us much of the oxygen we breathe—which is needed by all animal life. Other green plants give off oxygen, too. But most of it probably comes from the tiny plants of the sea.

Human beings and other animals take in oxygen and breathe out carbon dioxide

The most common plant plankton are the diatoms. Each
is a single cell, or bit of living matter, enclosed in a box
of silica, which is like glass. Diatoms have been called
"the grass of the sea," because so many other creatures feed
on them. They are important for this reason, as well as
for making oxygen.

Diatoms of the sea

Diatoms of fresh water

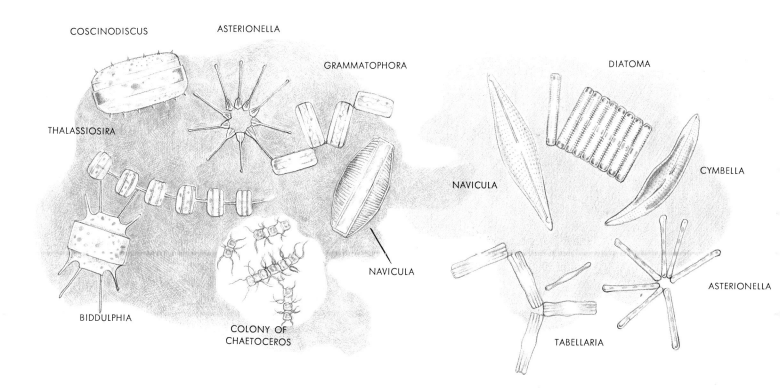

COSCINODISCUS

ASTERIONELLA

GRAMMATOPHORA

DIATOMA

THALASSIOSIRA

NAVICULA

CYMBELLA

BIDDULPHIA

COLONY OF
CHAETOCEROS

NAVICULA

TABELLARIA

ASTERIONELLA

There are other green plankton that act something like plants and something like animals. These can make food with chlorophyll and also catch and eat other plankton. They are called flagellates. Each has at least one flagellum, a whiplike "tail" that helps it swim upward toward the sunlight.

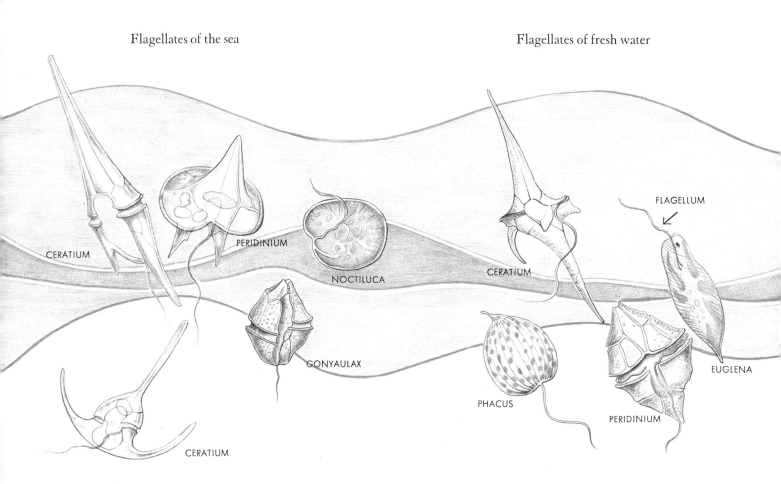

Flagellates of the sea

Flagellates of fresh water

CERATIUM

PERIDINIUM

NOCTILUCA

GONYAULAX

CERATIUM

FLAGELLUM

CERATIUM

PHACUS

PERIDINIUM

EUGLENA

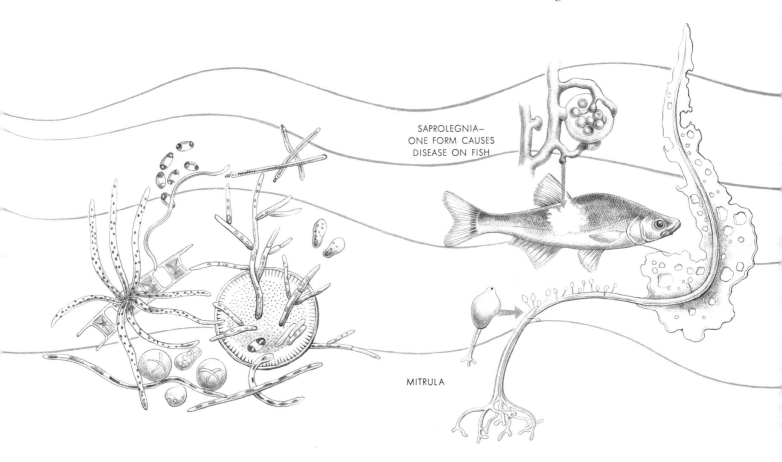

SAPROLEGNIA—
ONE FORM CAUSES
DISEASE ON FISH

MITRULA

The smallest and least known of the plant plankton are the bacteria and fungi. They are not green. Most of them take in their food from dead plants and animals in the water. They change dead things by making them rot away into useful chemicals that are "fertilizer" for green plants.

How a one-celled plant reproduces—
one diatom turning into two

The tiny plants of the plankton have a growing season
much like plants on land. In winter, sunlight is weak.
The plants' food-making slows down. Storms stir up the
water and mix fertilizing chemicals through it. In spring,
the sunlight becomes strong and the water warms up.
Plankton plants multiply quickly by splitting in two.
Their sudden growth is called a "bloom."

A part of the sea may have such a heavy bloom of flagellates that the water looks colored. People call it a "red tide" or "yellow tide." Fishes, crabs, shellfish, and other animals caught in this water can be suffocated when plankton clog their breathing organs. And dying flagellates can also poison the water and cause fishes and crabs to die by the millions.

GONYAULAX

EXUVIELLA

GONYAULAX

Poisonous
flagellates

In summer, warm water gathers in a layer near the surface. The plants in it use up the fertilizing chemicals, then stop growing. But water pollution from factories, sewage plants, and detergents may add extra chemicals to the water.

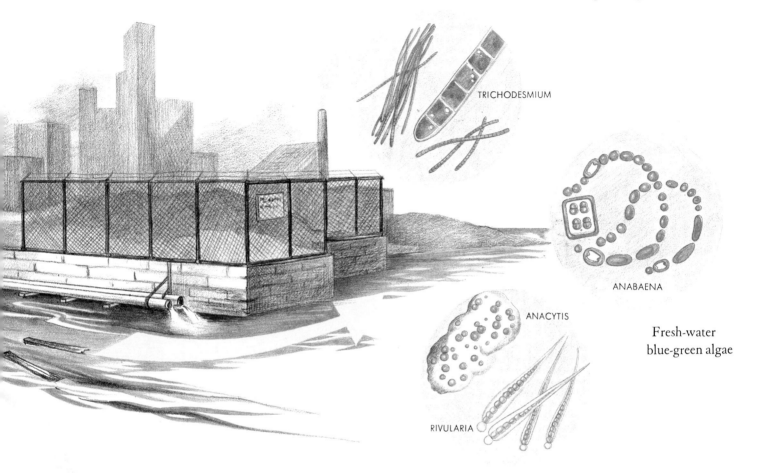

Marine blue-green algae

TRICHODESMIUM

ANABAENA

Fresh-water
blue-green algae

ANACYTIS

RIVULARIA

Tiny plants called blue-green algae multiply greatly in
polluted water. They can make water smell and taste
bad, and even poison other water creatures. This kind
of deadly bloom occurs most often in fresh water. But
it can happen off seaside cities, too.

When plankton bloom naturally, it means more food for the little animals that feed on it. They can multiply, too. Parts of the ocean with much plant plankton have many animal plankton also. But where plants are scarce, the waters are a wet "desert." The seas are not evenly full of life.

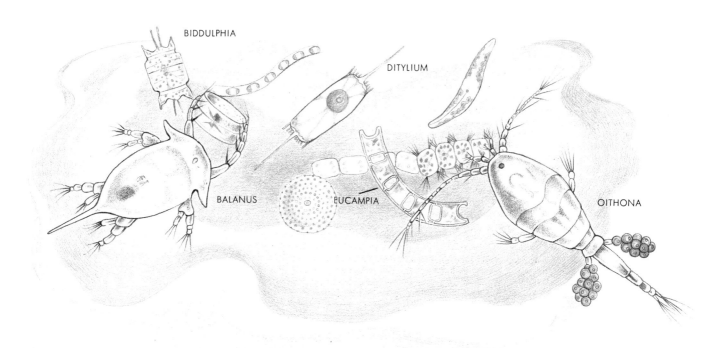

BIDDULPHIA

DITYLIUM

BALANUS

EUCAMPIA

OITHONA

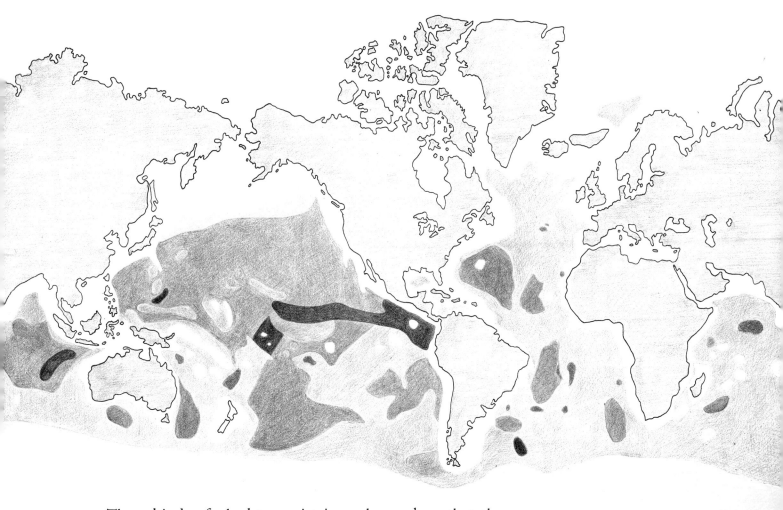

Three kinds of plankton exist in such numbers that they help to build the sea bottom. They are the diatoms, and one-celled animals called foraminiferans and radiolarians. All three have hard skeletons that drift to the bottom when the creatures die. These remains form thick beds of mud, called ooze.

 YELLOW: DIATOMS

 PURPLE: RADIOLARIANS

 GREEN: PTEROPODS

 BLUE: FORAMINIFERANS

 RED: RED CLAY

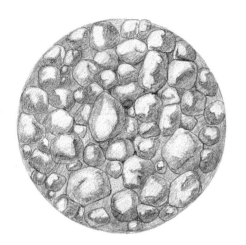

MUD FROM THE EARTH

Ooze covers tens of thousands of square miles of sea floor. Some ancient ooze beds became dry land and are now rock. Scientists think the oily liquid called petroleum may have been made from the remains of certain diatoms that multiplied greatly, then suddenly died. If this is true, then we get our oil and gasoline from plankton that lived millions of years ago.

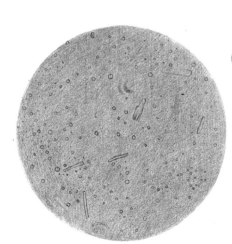

RED CLAY

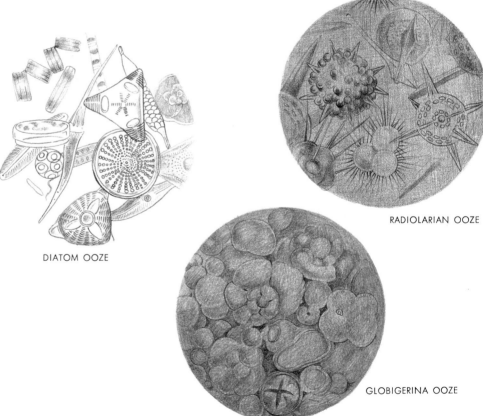

DIATOM OOZE

RADIOLARIAN OOZE

GLOBIGERINA OOZE

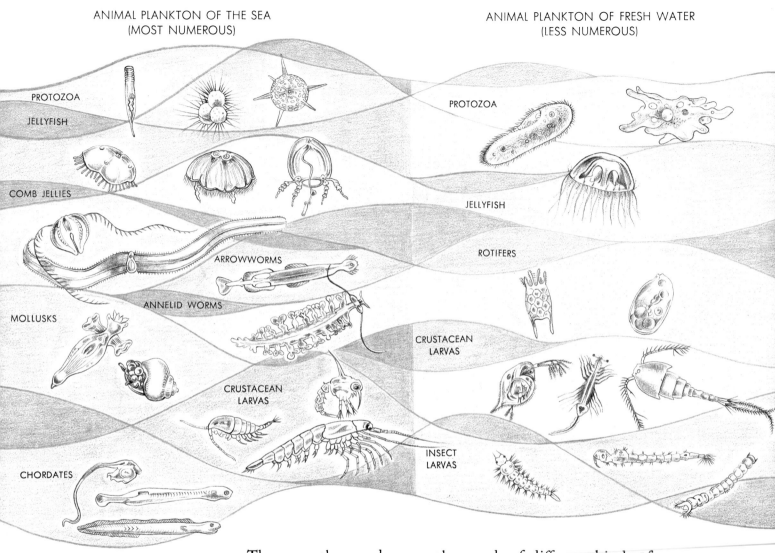

ANIMAL PLANKTON OF THE SEA
(MOST NUMEROUS)

ANIMAL PLANKTON OF FRESH WATER
(LESS NUMEROUS)

PROTOZOA

JELLYFISH

COMB JELLIES

ARROWWORMS

ANNELID WORMS

MOLLUSKS

CRUSTACEAN LARVAS

CHORDATES

PROTOZOA

JELLYFISH

ROTIFERS

CRUSTACEAN LARVAS

INSECT LARVAS

There are thousands upon thousands of different kinds of
plankton animals, belonging to all branches of the animal
kingdom. Here are a few of the large groups, with some
common plankton belonging to them.

The most active plankton animals are the predators, which chase or snare their food. Jellyfish have poison darts that paralyze animal prey. Comb jellies "lasso" little creatures with trailing threads. Arrowworms open their bristling jaws, zip about like glass torpedoes and catch animals as large as themselves.

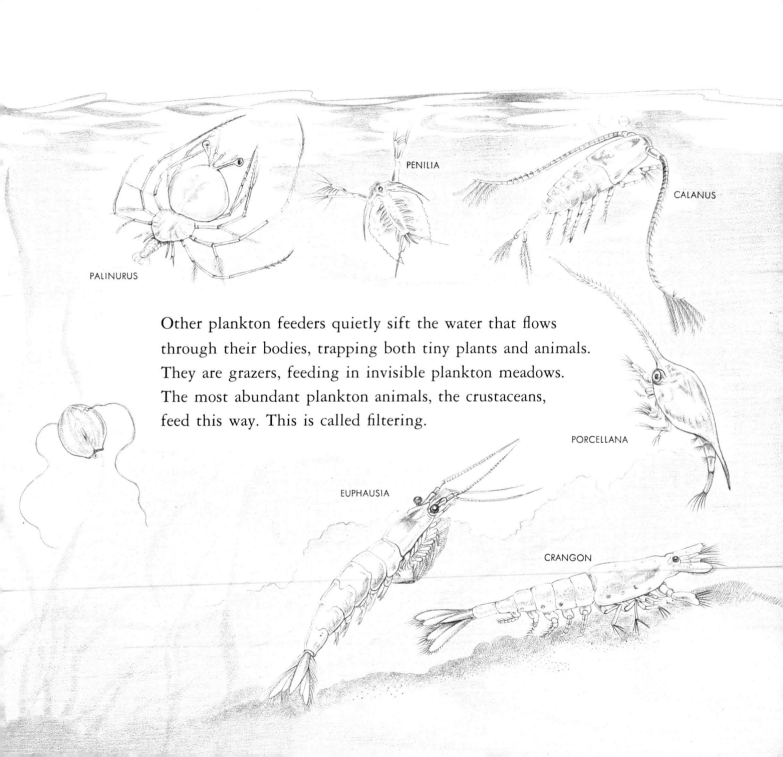

PALINURUS

PENILIA

CALANUS

PORCELLANA

Other plankton feeders quietly sift the water that flows
through their bodies, trapping both tiny plants and animals.
They are grazers, feeding in invisible plankton meadows.
The most abundant plankton animals, the crustaceans,
feed this way. This is called filtering.

EUPHAUSIA

CRANGON

Krill are two-inch crustaceans like little shrimp. They feed on diatoms and other tiny plants. In some parts of the ocean — as off western South America — deep waters flow up to the surface, bringing rich chemicals that make the diatoms grow. The krill are present there in such vast numbers that they are food for huge whales.

BALEEN
(STRAINER)

WHALE JAW

KRILL
ENLARGED

EUPHAUSIA

Many other creatures, large and small, feed on plankton animals in the sea. The giant basking shark, penguins, and crab-eater seals eat mostly krill. Sardines, anchovies, and herring — valuable food for man — eat other tiny crustaceans. Clams, oysters, mussels, and other shellfish filter the waters of smaller drifting life.

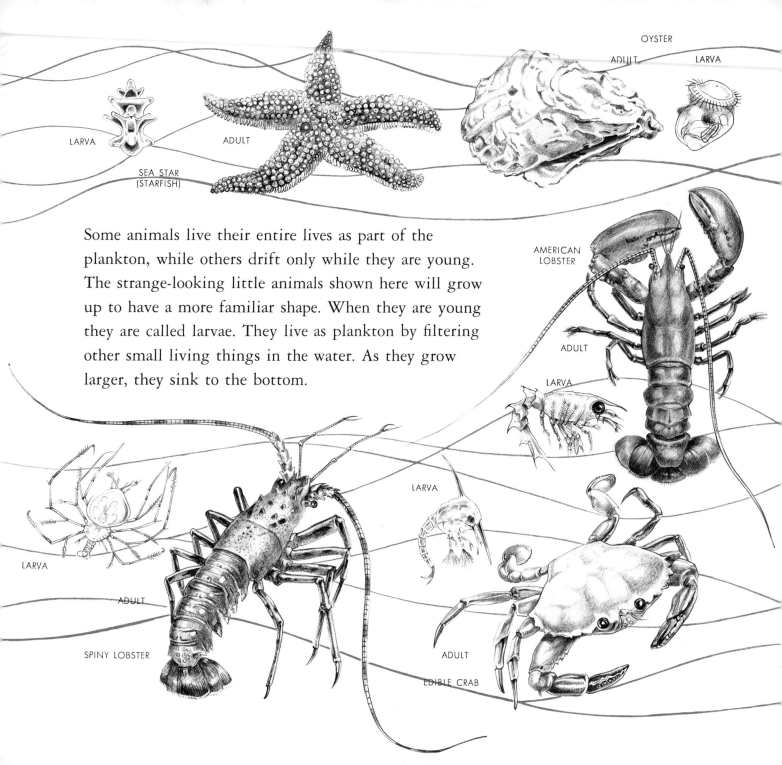

LARVA

SEA STAR
(STARFISH)

ADULT

OYSTER

ADULT          LARVA

AMERICAN
LOBSTER

ADULT

LARVA

Some animals live their entire lives as part of the
plankton, while others drift only while they are young.
The strange-looking little animals shown here will grow
up to have a more familiar shape. When they are young
they are called larvae. They live as plankton by filtering
other small living things in the water. As they grow
larger, they sink to the bottom.

LARVA

LARVA

ADULT

SPINY LOBSTER

ADULT

EDIBLE CRAB

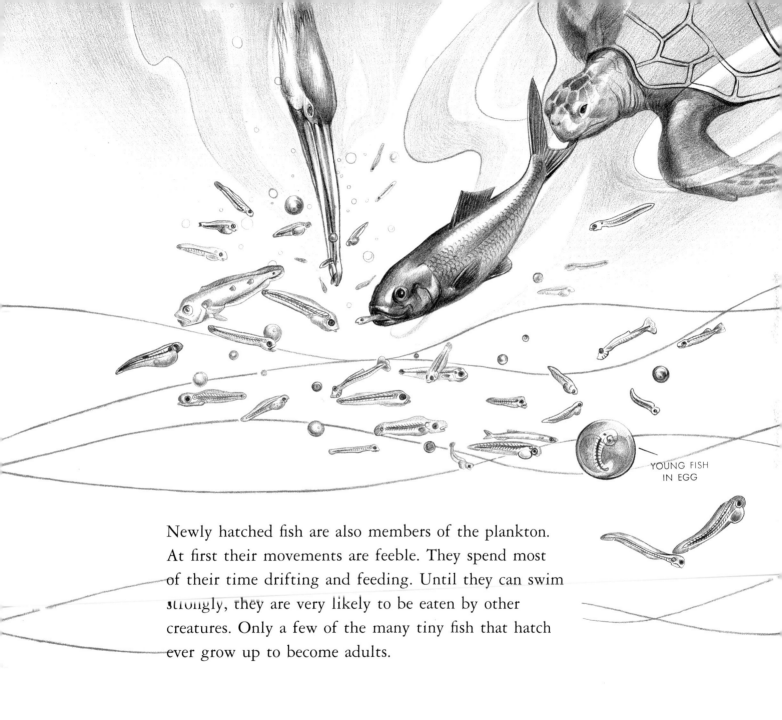

YOUNG FISH
IN EGG

Newly hatched fish are also members of the plankton.
At first their movements are feeble. They spend most
of their time drifting and feeding. Until they can swim
strongly, they are very likely to be eaten by other
creatures. Only a few of the many tiny fish that hatch
ever grow up to become adults.

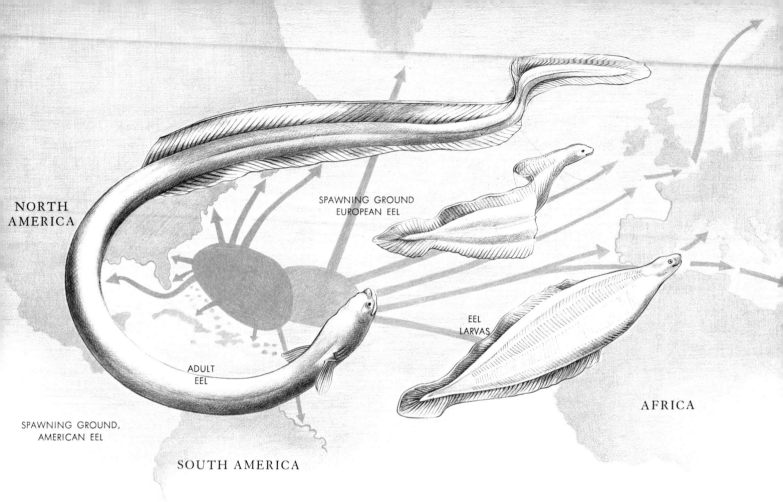

NORTH
AMERICA

SPAWNING GROUND
EUROPEAN EEL

EEL
LARVAS

ADULT
EEL

AFRICA

SPAWNING GROUND,
AMERICAN EEL

SOUTH AMERICA

One of the most famous young fish of the plankton is the eel larva. It looks like a glass leaf — not at all like its parents. Adult eels live in rivers of Europe and North America. They swim to the sea, then travel to the mid-Atlantic to reproduce. The young, carried by water currents, slowly drift back to the shores that their parents came from.

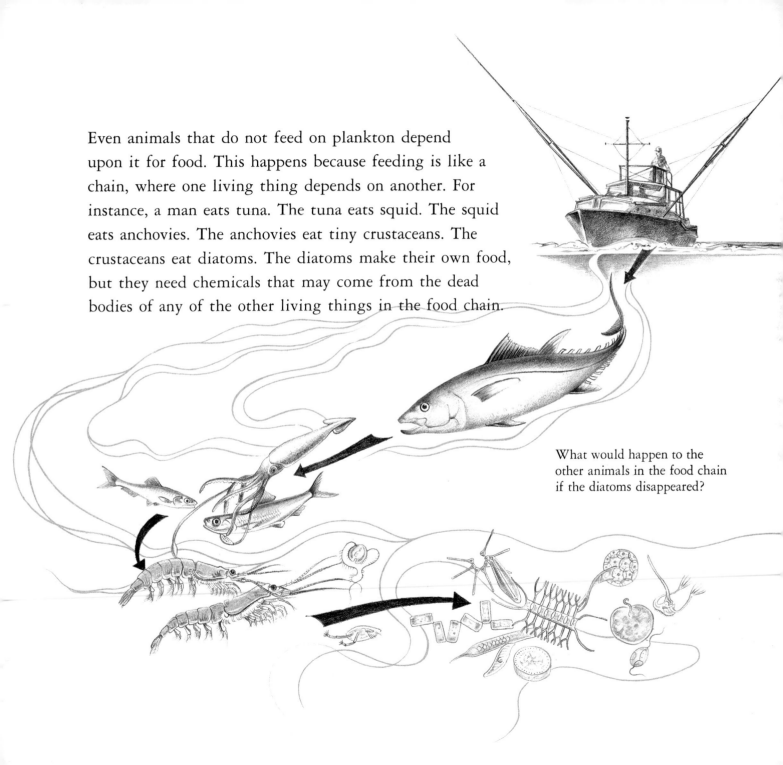

Even animals that do not feed on plankton depend upon it for food. This happens because feeding is like a chain, where one living thing depends on another. For instance, a man eats tuna. The tuna eats squid. The squid eats anchovies. The anchovies eat tiny crustaceans. The crustaceans eat diatoms. The diatoms make their own food, but they need chemicals that may come from the dead bodies of any of the other living things in the food chain.

What would happen to the other animals in the food chain if the diatoms disappeared?

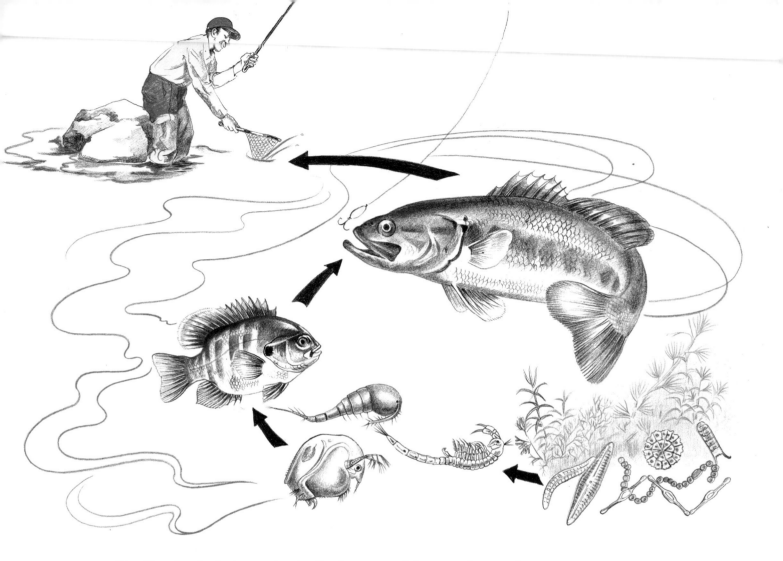

Similar food chains exist in fresh water. Man catches and
eats largemouth bass. The bass eats sunfish. The sunfish eat
copepods, water fleas, and other tiny crustaceans. Crustaceans
eat plant plankton, and the plants thrive because of the
chemicals in the water.

Drawings of young and adult acorn barnacle
by J. Vaughan Thompson, 1828

Charles Darwin's "Beagle"
using a bottom dredge,
1845

Until the early 1800s, men hardly knew that plankton existed. And it is only in recent years that marine scientists have discovered how important it is. Today, research ships collect and study plankton with special equipment. Sound waves can also be bounced off plankton to show how they move up and down.

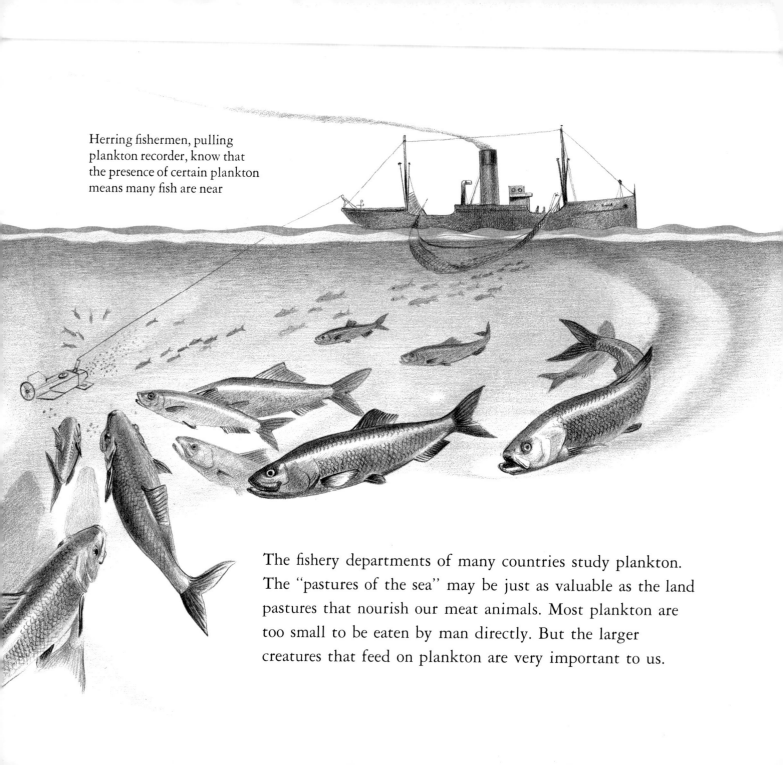

Herring fishermen, pulling plankton recorder, know that the presence of certain plankton means many fish are near

The fishery departments of many countries study plankton. The "pastures of the sea" may be just as valuable as the land pastures that nourish our meat animals. Most plankton are too small to be eaten by man directly. But the larger creatures that feed on plankton are very important to us.

Scientists want to learn why plankton plants bloom and suddenly die, killing fish and other animals. They want to know how man's pollution of the waters is affecting these tiny living things. They have already warned us that some lakes and ocean bays are "dying" because pollution has upset the natural food chain in the waters.

Most of the tiny life of the waters can't be seen by us. It lives and multiplies endlessly under the sun. Then it dies and is made into chemicals that build new life. The plankton give us oxygen to breathe and vast amounts of food. They even give us hints of what life was like billions of years ago, when the first living things were plankton, floating quietly in the sea.

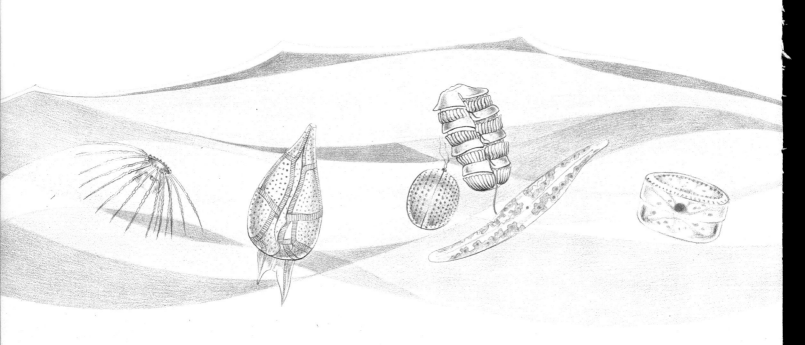